Beginning Algebra Inquiry Activities

Activities to use when teaching the basic concepts of Pre-Algebra and Algebra!

For Grades 6 through 10

Claire Hubacher Johnson

©2006

J&B Products, Inc.
P.O. Box 3010, Janesville, WI 53547-3010

Printed in the United States of America

ISBN 0-9776782-1-0

Beginning Algebra Inquiry Activities

Table of Contents

Introduction/Instructions	3
Variable Expressions	13
Coordinate Plane	17
Graphing	21
Solving Equations & Inequalities	25
Fractions	29
Factors & Exponents	33
Student Worksheets	36
Transparencies	Pocket

All rights reserved. The activities in this book are provided to teach and extend learning experiences. These are copyright protected and are reproducible for educational purposes only.

$6(9 - 7) + x > 15$

Beginning Algebra Inquiry Activities

Activities and games are great for adding interest and fun to lesson plans! These activities work well for supplementing general text assignments or use them for introducing topics, pre-tests, review, practice, quizzes and reinforcement. Versatile enough to use with individuals, small-groups, or large-groups.

Using This Book:

Beginning Algebra Inquiry Activities covers the following topics:

◊ **Variable Expressions**
◊ **Coordinate Plane**
◊ **Graphing**
◊ **Solving Equations & Inequalities**
◊ **Fractions**
◊ **Factors & Exponents**

For each of the activities/games listed below, you can use a single category, several categories, or all of the categories and questions. Divide the class or group into teams, or if working with a small group, a "team" can be just one individual. When playing in teams, encourage players to consult each other before deciding on the answer. You may want to set time limits for responding to the questions.

There are a variety of ways to use these questions. Choose your favorite game version listed in these instructions and review the "Curriculum Uses" section for additional activity ideas. Motivate students by using the scores as extra credit points. Divide the total team points evenly amongst team members and add to their grades.

Depending on the skill level of your students, you may want to have students use pencil and paper for determining answers or use mental math instead. Having students show their written work and verbalize their work are added dimensions that can be used to determine whether or not the students comprehend what they are learning.

Basic Activity:

Determine the category that will be used. If this topic is new to the players, discuss/teach the topic information presented before doing this activity. Place the corresponding transparency on the overhead projector or provide as a handout if working with an individual. The transparencies are a list of the answers. Some transparencies include extra answers that won't be used at all, while some answers will be used more than once.

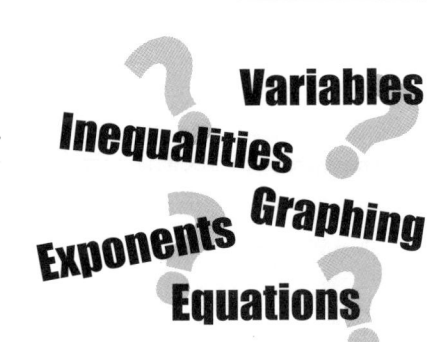

Choose which team will go first. Pick any question in the category and read it aloud to the first team. The team responds by choosing the correct answer listed on the transparency.

If the team responds correctly, they get the number of points listed on the question, and receive a bonus question that only they can answer. Only one bonus question is given for each turn. If the team responds incorrectly, they either do not receive any points, or for greater difficulty, the points can be deducted from the team's score. The game ends when time is called. The winning team is the one with the most points. Keep track of points on the whiteboard.

Variations for Basic Activity:

Team Responses: Instead of asking the question to a particular team, all teams are eligible to respond. In order to answer, they must be the first team to raise their hand, jump to their feet, blow a whistle, or use some other predetermined method to show they were first.

Stealing the Question: If a team does not answer their question correctly, the other team(s) can steal by providing the correct response and then winning the points.

Raising the Stakes: After responding correctly, the team can choose to either keep going and receive another question, or stop and keep the points. If they keep going, and answer the next question correctly, they will receive the total points for both questions. If they respond incorrectly, they will lose the points earned on the first, (second and third,) question as well, and play will pass to the next team. Teams can keep going and respond correctly to as many questions as they choose in a round, or you can set a limit such as 3 questions for each round. When they choose to stop, (after responding correctly to a question,) they receive the total of all points earned in that round.

Advanced Activity: Do not use the answer transparency. Instead, players must solve the problems and determine the answers themselves.

Baseball:

Draw a baseball diamond on the whiteboard or use the overhead transparency provided in this book. When using both, the baseball and answer transparencies, you will need to flip between the two transparencies. Otherwise, the answer transparency could be printed and distributed to students, while the "game" transparency is used on the overhead. Duplicate the question sheets and cut them apart so that each question is on a separate piece of paper. Mix up the questions and stack them face down. Choose a scorekeeper and game leader/reader. Divide the class/group into two teams. The point values listed on the questions will be used to determine the scoring:

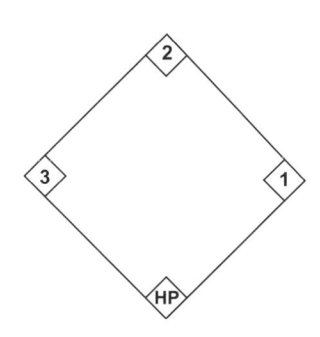

    5 points = a single   10 points = a double
    15 points = a triple   20 points = a home run
    25 points = a grand slam (A grand slam is a home run with bases loaded = 4 runs scored.)

The leader asks a question to the first team up to bat. If their answer is correct, they move to the appropriate base that matches the point value of the question. Use different colored erasable markers to keep track of movement around the bases. If they answer incorrectly, they are out! Continue to ask questions of the same team until they get 3 outs. (You may want to limit the number of questions asked in each inning in order to keep everyone involved and active.) Keep track of the runs scored as questions are correctly answered. The winner is the team with the highest score at the end of the game. Play as many innings as desired.

Variation for Baseball: Instead of drawing a diamond on the whiteboard, or using the transparency, move the desks around in the classroom and set-up "bases" for team members to move to when they correctly answer a question.

Tic-Tac-Toe:

This is a quick and easy game. Divide the class/group into two teams and choose which team is the "X" team and which one is the "O" team. Choose which team goes first. Draw a Tic-Tac-Toe board on the whiteboard or use the overhead transparency provided in this book. When using both, the Tic-Tac-Toe and answer transparencies, you will need to flip between the two transparencies. Otherwise, the answer transparency could be printed and distributed to students, while the "game" transparency is used on the overhead. The point values listed on the questions will not be used for this game.

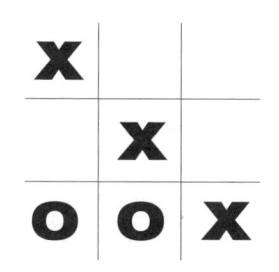

The leader asks a question to the first team. If they respond correctly, they make their mark on the Tic-Tac-Toe board. If they respond incorrectly, they do nothing. Play then moves to the other team and continues in the same manner. The first team to make Tic-Tac-Toe is the winner. Great for use when only a small amount of time is available. Spending 10 minutes every day playing Tic-Tac-Toe is a great review activity.

Essential Facts:

Draw an Essential Facts chart on the whiteboard or use the overhead transparency included in this book. When using the Essential Facts transparency with the answer transparency, you will need to flip between the two transparencies. Otherwise, the answer transparency could be printed and distributed to students, while the "game" transparency is used on the overhead.

	Category	Category	Category	Category	Category	Category
5						
10						
15						
20						
25						

FIGURE 1

When playing with all categories at once, (see figure 1) write in the category names across the top row of the playing chart, and choose one question for each point value in each category. If playing with only one category, (see figure 2) you will use only 3 columns of the playing chart, so write A, B, and C, as your category names. You will then use all of the questions given for that category. (There are three questions for each point value in each category.)

	A	B	C
5			
10			
15			
20			
25			

FIGURE 2

Choose which team goes first. They then choose a category and point value. The game leader reads the question. All teams have a chance to answer the question by being the first team to raise their hands, jump to their feet, blow a whistle, or use another method to show they were first. If they answer correctly, they earn the point value of that question and select the next question and point value.

If they respond incorrectly, another team can steal by giving the correct answer and earn the points; or the leader can read the correct response, and the team with the previous correct answer chooses the next question.

Check off the used question on your list so it doesn't get used again during that round, and place an "X" on the transparency for that category so that players know they can no longer select that point value. If desired, the point value of incorrect answers may be deducted from scores. Play continues until all cards are used or time has expired. The team scoring the most points is the winner.

Variations for Essential Facts:

Marathon Play: Use all 90 questions (15 questions for each category), and continue playing until all questions are used or class time has expired.

Bonus Round: Double the point values. Limit this version to the final 5-8 minutes of play. Only the first team to respond has the opportunity to win the extra points.

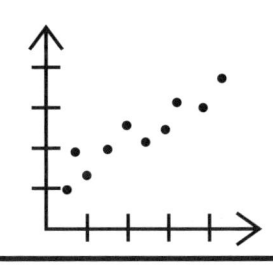

Curriculum Uses for Beginning Algebra Inquiry Activities

Worksheets/Study Guides:

Duplicate the student worksheets provided in this packet and give one to each student. Space is provided on the worksheets for students to write in their answers. Use these worksheets in class for taking notes during the lesson or students can complete them as an individual take-home assignment.

Quizzes or Tests:

The 90 questions are great for quizzes, or tests. Either read the questions orally to students and have them write their answers on a sheet of paper, or use the worksheets provided in this book as test papers.

Application/Extra Credit:

Have students provide real-life examples of these topics. Examples can be found in architectural designs, interior decorating, landscaping, general building construction, cooking, medicine, geology, etc. Possible examples include:

1. Calculate the cost of mowing the lawn when given a per sq. ft. rate.
2. Calculate how much fertilizer will be needed to cover your backyard lawn.
3. Determine longitude and latitude for various places on a map. Compare with known weather data for a specific latitude and be a weatherman by drawing conclusions that predict weather patterns for a region.
4. Research vacation trips on the internet to determine what season would be the cheapest time to travel to a particular destination.
5. Map out a road trip using several routes and determine which route is shortest.
6. Provide a target heart rate and determine what the resting heart rate would be or the ideal heart rate for a cardiovascular workout.
7. Provide a unit of measure for medicine at a particular weight and be the doctor that must determine the unit of measure at a larger and smaller weight.
8. Graph population data for a community over a certain period of time. Be a community planner and draw conclusions from the data to determine growth patterns and future needs based on the data, such as will school population increase, are more senior residence facilities needed, etc.

Note: A graph transparency and copy master for graph paper have been provided on pages 11 and 12 for use with students as needed.

Beginning Algebra Inquiry Activities:
Baseball Diamond Transparency

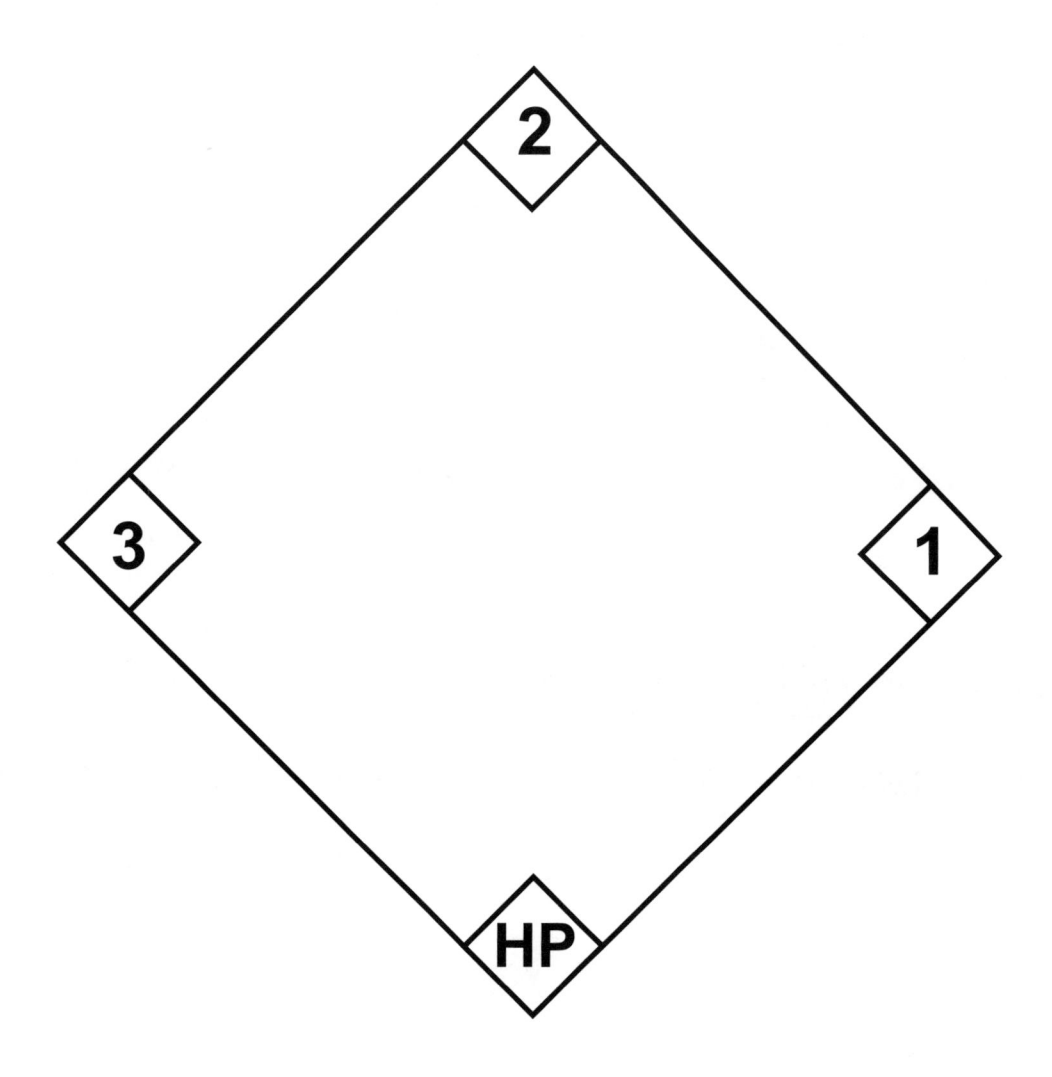

Beginning Algebra Inquiry Activities:
Tic Tac Toe Transparency

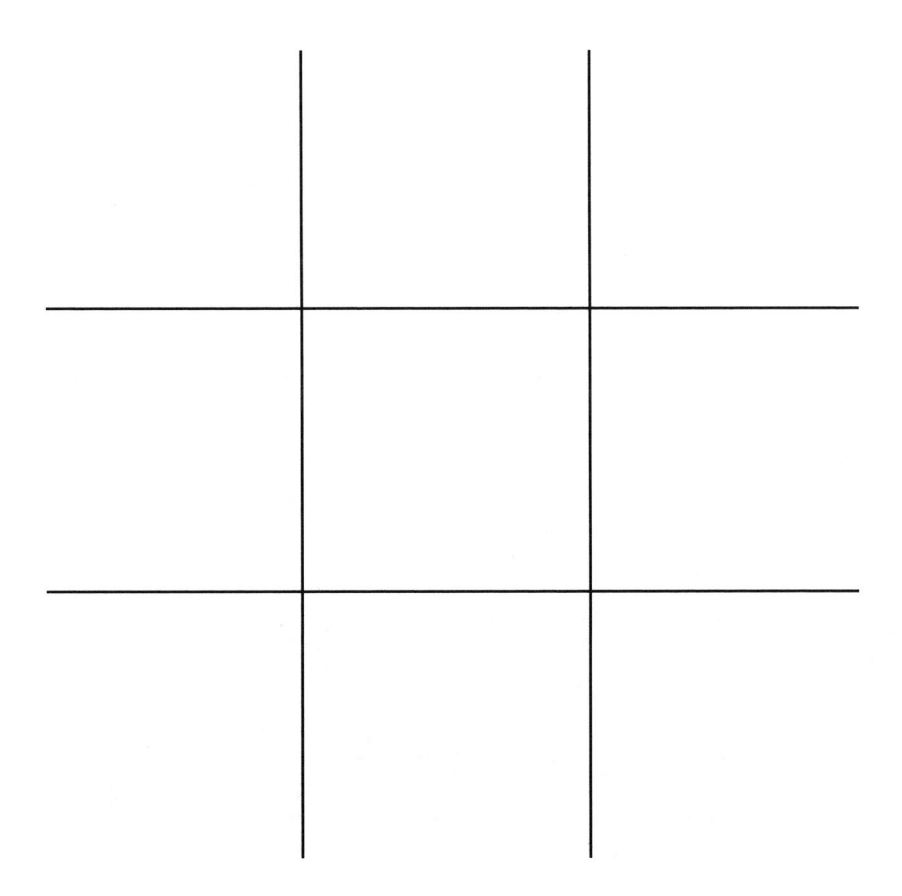

Beginning Algebra Inquiry Activities: Essential Facts Transparency

	Category	Category	Category	Category	Category	Category
5						
10						
15						
20						
25						

	A	B	C
5			
10			
15			
20			
25			

©2006 ~ J&B Products, Inc., P.O. Box 3010, Janesville, WI 53547

Beginning Algebra Inquiry Activities:
Graph Transparency

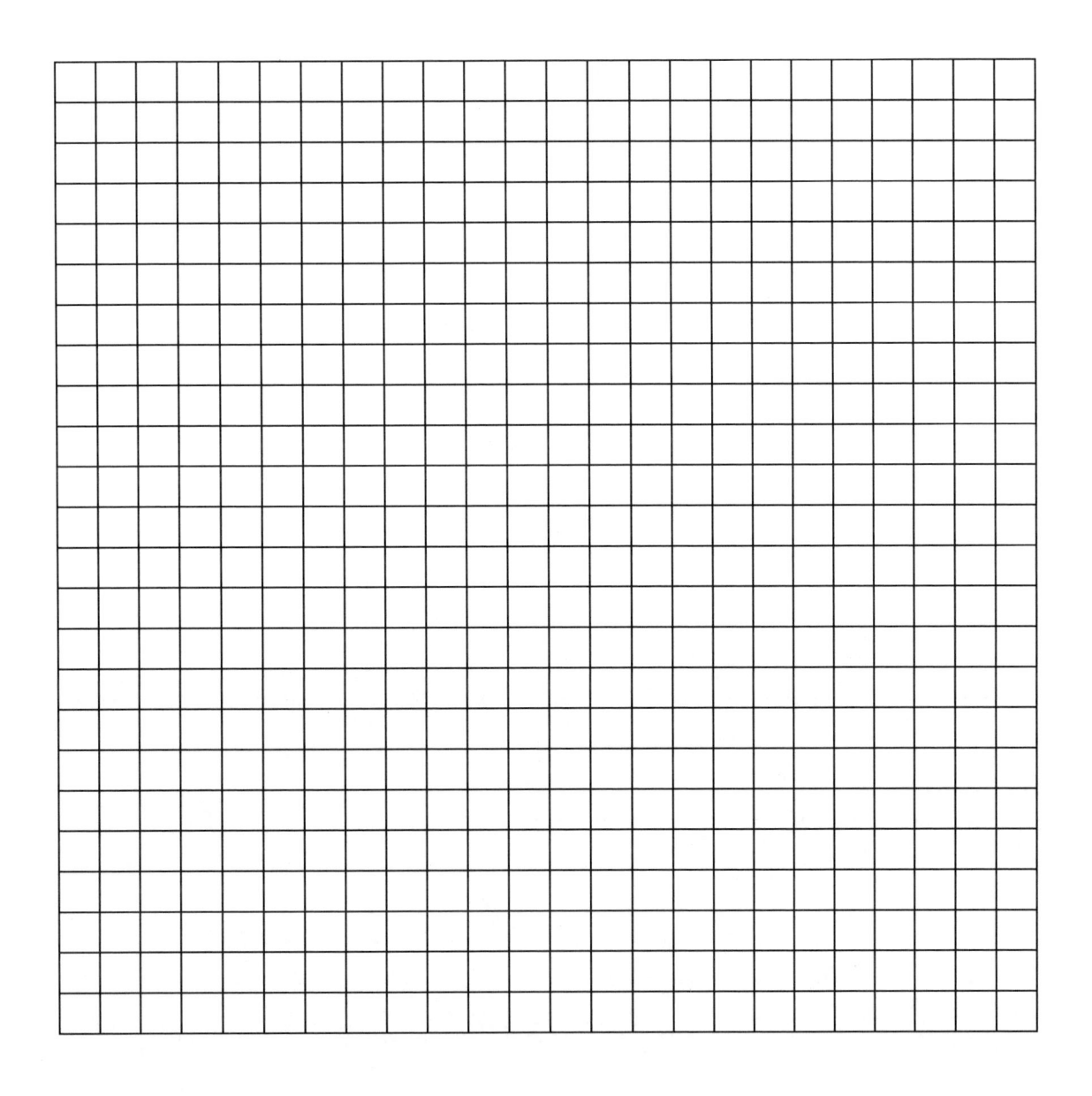

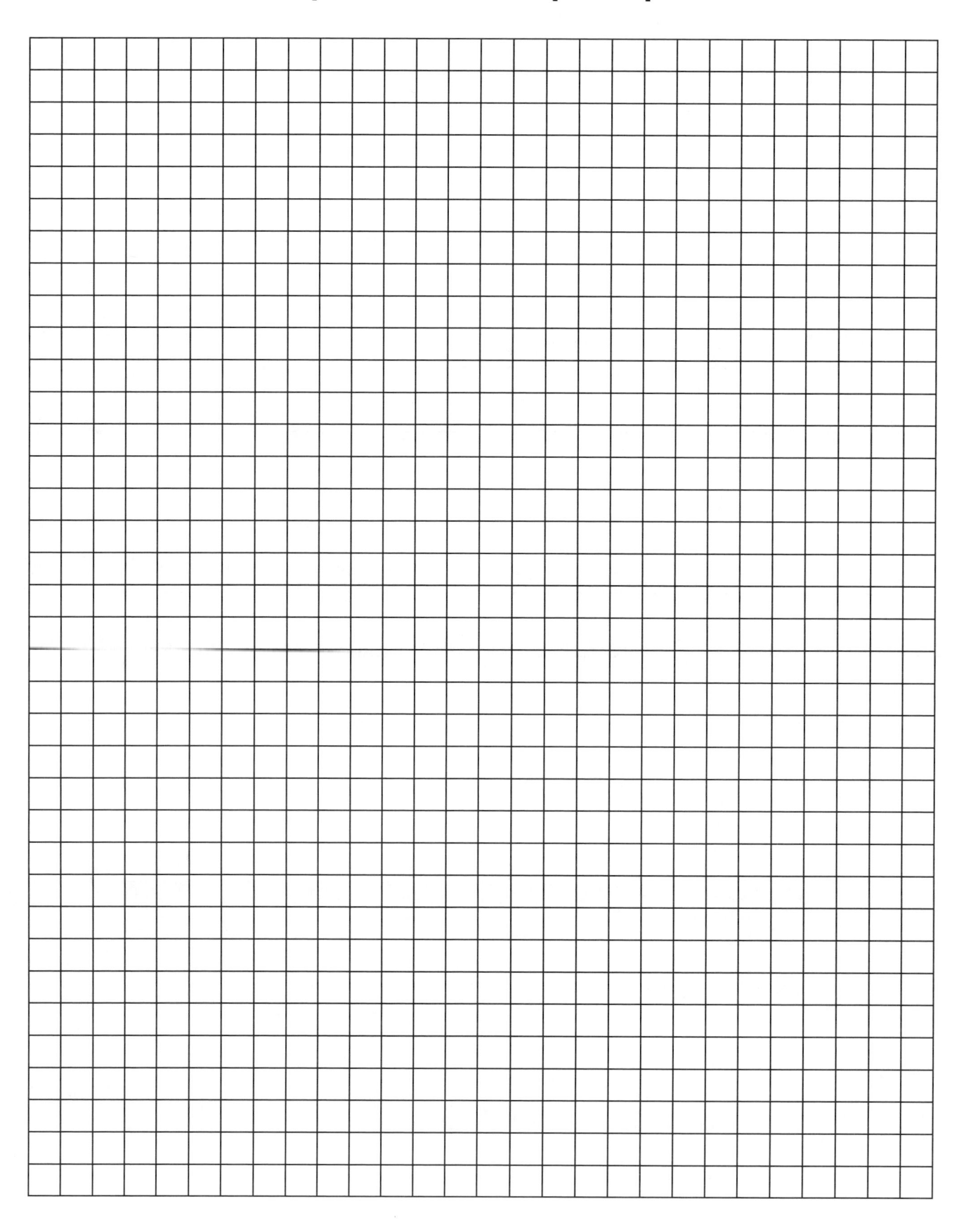

Beginning Algebra Inquiry Activities:
Reproducible Graph Paper

Beginning Algebra Inquiry Activities: Teacher's Key

Variable Expressions: The transparency includes extra answers that will not be used.

1. What is the variable expression for "7 less than x"?
   (Point Value: 5)

   Transparency Answer E; x–7

2. Simplify: 15 + 20 • 3
   (Point Value: 5)

   Transparency Answer J; 75

3. Using inequality or equality symbols, compare 0 ▓ -100.
   (Point Value: 5)

   Transparency Answer A; >

4. Simplify: -48 ÷ 2
   (Point Value: 10)

   Transparency Answer C; -24

5. What is the variable expression for the "product of 7 and n"?
   (Point Value: 10)

   Transparency Answer O; 7n

6. Simplify: -16 + -2
   (Point Value: 10)

   Transparency Answer I; -18

7. Using inequality or equality symbols, compare -12 ▓ -9.
   (Point Value: 15)

   Transparency Answer G; <

13

8. Simplify: 9 • 8 − 12 ÷ 2
   (Point Value: 15)

   Transparency Answer K; 66

9. Simplify -1(-2)(-3)(-4)
   (Point Value: 15)

   Transparency Answer D; 24

10. Simplify: 5(8 - 4) ÷ 5 ÷ 2
    (Point Value: 20)

    Transparency Answer F; 2

11. What is the next number in this pattern 1, 1, 2, 3, 5, 8, 13, . . . ?
    (Point Value: 20)

    Transparency Answer N; 21

12. What is the variable expression for "7 times the quantity 4 plus x"?
    (Point Value: 20)

    Transparency Answer M; 7(4 + x)

13. Simplify: 14 − -4
    (Point Value: 25)

    Transparency Answer B; 18

14. Using inequality or equality symbols, compare |-4| |4|.
    (Point Value: 25)

    Transparency Answer L; =

15. Solve: 15x − 2x, for x=3
    (Point Value: 25)

    Transparency Answer R; 39

Beginning Algebra Inquiry Activities: Variable Expressions Transparency

A. >   B. 18   C. -24

D. 24   E. x-7   F. 2

G. <   H. -39   I. -18

J. 75   K. 66   L. =

M. 7(4+x)   N. 21   O. 7n

P. -66   Q. 22   R. 39

Beginning Algebra Inquiry Activities: Teacher's Key

Coordinate Plane: Students will need graph paper for some of these problems. The transparency includes extra answers that will not be used.

1. What is formed when two number lines intersect at their zero points?
   (Point Value: 5)

   Transparency Answer F; Coordinate plane

2. What is the term for two numbers that tell the location of a point in a coordinate plane?
   (Point Value: 5)

   Transparency Answer M; Ordered pair

3. In a coordinate pair, which axis is referenced first?
   (Point Value: 5)

   Transparency Answer A; x-axis

4. In which quadrant, are the coordinates both positive?
   (Point Value: 10)

   Transparency Answer L; Quadrant I

5. What is the number of the quadrant for an ordered pair with two negative numbers?
   (Point Value: 10)

   Transparency Answer E; Quadrant III

6. What is the quadrant for an ordered pair when the first number is positive and the second number is negative?
   (Point Value: 10)

   Transparency Answer P; Quadrant IV

7. Connect the points in the order given. What is the shape of (2, 4), (2, -2), (-5, -2), (-5, 4)?
   (Point Value: 15)

   Transparency Answer H; Rectangle

8. Connect the points in the order given. What is the shape of (2, -4), (7, -1), (4, 4), (-1, 1)?
   (Point Value: 15)

   Transparency Answer B; Square

9. Which quadrant would (-2, 1) be located in?
   (Point Value: 15)

   Transparency Answer C; Quadrant II

10. Which coordinate shows if the position is above or below the x-axis?
    (Point Value: 20)

    Transparency Answer N; y-coordinate

11. Connect the points in the order given. What is the shape of (-1, 2), (1, 5), (7, 5), (5, 2)?
    (Point Value: 20)

    Transparency Answer J; Parallelogram

12. Which ordered pair would be in Quadrant I?
    (Point Value: 20)

    Transparency Answer G; (5, 4)

13. Which two ordered pairs make a line that is parallel to the x-axis?
    (Point Value: 25)

    Transparency Answer R; (-3, 5), (3, 5)

14. Which two ordered pairs make a line that is parallel to the y-axis?
    (Point Value: 25)

    Transparency Answer K; (-3, 5), (-3, -5)

15. Which ordered pair would be in Quadrant IV?
    (Point Value: 25)

    Transparency Answer D; (5, -4)

Beginning Algebra Inquiry Activities: Coordinate Plane Transparency

A. x-axis

B. Square

C. Quadrant II

D. (5, -4)

E. Quadrant III

F. Coordinate plane

G. (5, 4)

H. Rectangle

I. Trapezoid

J. Parallelogram

K. (-3, 5), (-3, -5)

L. Quadrant I

M. Ordered pair

N. y-coordinate

O. (-3, -4)

P. Quadrant IV

Q. (-3, 4)

R. (-3, 5), (3, 5)

Beginning Algebra Inquiry Activities: Teacher's Key

Graphing: Students will need graph paper for some of these problems. Some graphs on the transparency will be used more than once. The transparency also includes extra answers that will not be used.

1. Which graph represents x = 2?
   (Point Value: 5)

   Transparency Answer Graph A

2. Which graph represents y = 2?
   (Point Value: 5)

   Transparency Answer Graph B

3. Which graph shows no correlation?
   (Point Value: 5)

   Transparency Answer Graph E

4. Which graph shows a line whose slope is 0?
   (Point Value: 10)

   Transparency Answer Graph B

5. Which graph shows a line whose slope is undefined?
   (Point Value: 10)

   Transparency Answer Graph A

6. Which graph demonstrates what happened to the weight of a group of students as the students increased in age?
   (Point Value: 10)

   Transparency Answer Graph C

7. Using the vertical-line test, which graph shows a relation that is not a function?
   (Point Value: 15)

   Transparency Answer Graph A

8. Which graph shows a positive correlation?
   (Point Value: 15)

   Transparency Answer Graph C

9. Which graph shows a negative correlation?
   (Point Value: 15)

   Transparency Answer Graph D

10. What is the slope of a line in the following equation: $y = 3x + b$?
    (Point Value: 20)

    Transparency Answer H; 3

11. The slope of a line is -2. Which two points does it pass through?
    (Point Value: 20)

    Transparency Answer L; (-1, 5), (2, -1)

12. What is the slope of a line in the following equation: $9x + 3y = 15$?
    (Point Value: 20)

    Transparency Answer K; -3

13. What is the slope of a line through the points (-6, 2), (-2, 4)?
    (Point Value: 25)

    Transparency Answer G; $\frac{1}{2}$

14. What is the slope of a line through the points (-2, 6), (4, 3)?
    (Point Value: 25)

    Transparency Answer J; $-\frac{1}{2}$

15. The slope of a line is 2. Which two points does it pass through?
    (Point Value: 25)

    Transparency Answer F; (2, 5), (-2, -3)

Beginning Algebra Inquiry Activities:
Graphing Transparency

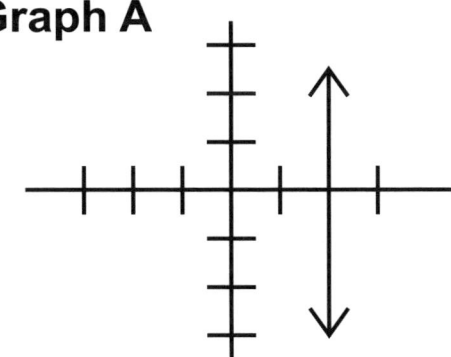

Graph A

Graph B

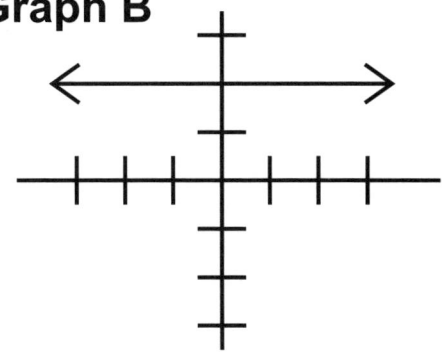

Graph C

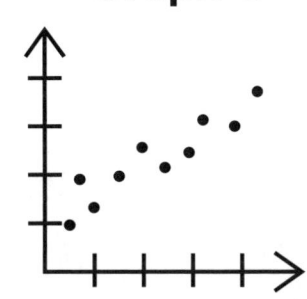

Graph D

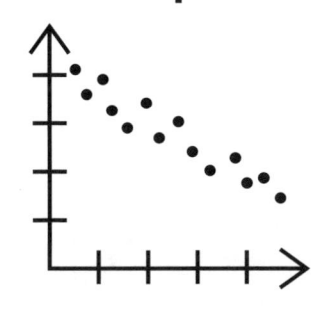

Graph E

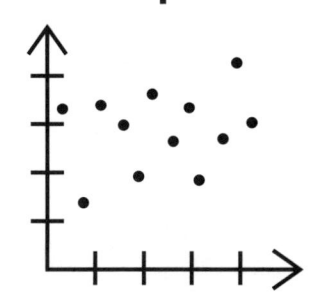

F. (2, 5), (-2, 3)

G. $\dfrac{1}{2}$

H. 3

I. $\dfrac{1}{3}$

J. $-\dfrac{1}{2}$

K. -3

L. (-1, 5), (2, -1)

M. $-\dfrac{1}{3}$

Beginning Algebra Inquiry Activities: Teacher's Key

Solving Equations & Inequalities: Some answers on the transparency will be used more than once.

1. Which equation shows the Commutative Property of Multiplication?
   (Point Value: 5)

   Transparency Answer H; ab = ba

2. Which equation shows the Distributive Property?
   (Point Value: 5)

   Transparency Answer F; a(b-c) = ab-ac

3. Which equation shows the Associative Property of Multiplication?
   (Point Value: 5)

   Transparency Answer K; (ab)c = a(bc)

4. Which graph shows y < 3?
   (Point Value: 10)

   Transparency Answer Graph B

5. Which graph shows y = 3?
   (Point Value: 10)

   Transparency Answer Graph A

6. Which graph shows y ≤ 3?
   (Point Value: 10)

   Transparency Answer Graph C

7. Solve: -3x + 6 = 18
   (Point Value: 15)

   Transparency Answer I; x = -4

8. Which graph shows y > 3?
   (Point Value: 15)

   Transparency Answer Graph D

9. Which graph shows y ≥ 3?
   (Point Value: 15)

   Transparency Answer Graph E

10. Solve: x + 1 ≤ 5
    (Point Value: 20)

    Transparency Answer G; x ≤ 4

11. Which graph shows the value of x in x(9 + 1) = 30?
    (Point Value: 20)

    Transparency Answer Graph A

12. Solve: $\frac{n}{2} - 8 = -6$

    (Point Value: 20)

    Transparency Answer L; n = 4

13. Which graph shows the value of x in 6(9 – 7) + x > 15?
    (Point Value: 25)

    Transparency Answer Graph D

14. Which graph shows the value of x in x – (9 – 2) ≤ -4?
    (Point Value: 25)

    Transparency Answer Graph C

15. Solve: 2y + 7 ≤ -1
    (Point Value: 25)

    Transparency Answer J; y ≤ -4

Beginning Algebra Inquiry Activities: Solving Equations & Inequalities Transparency

Graph A

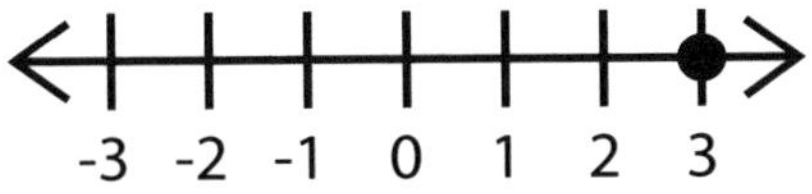

Graph B

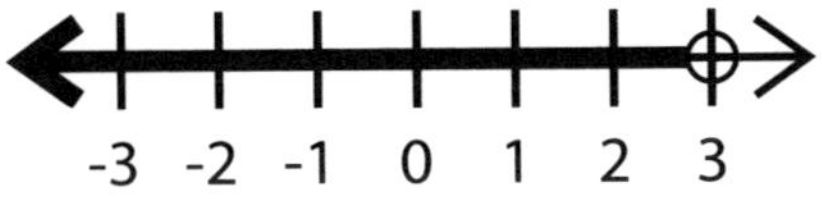

Graph C

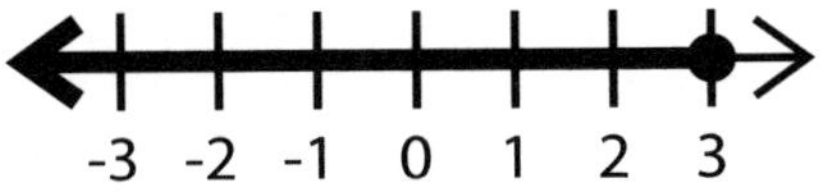

Graph D

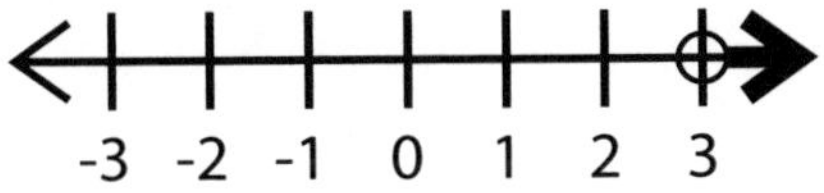

Graph E

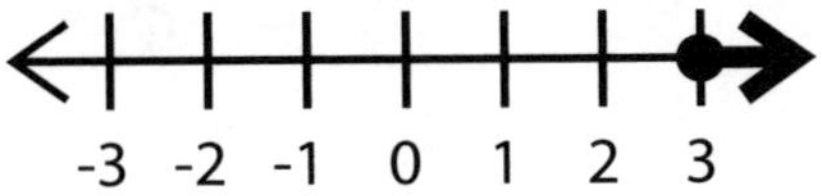

F. $a(b-c) = ab-ac$

G. $x \le 4$

H. $ab = ba$

I. $x = -4$

J. $y \le -4$

K. $(ab)c = a(bc)$

L. $n = 4$

Beginning Algebra Inquiry Activities: Teacher's Key

Fractions: Some answers on the transparency will be used more than once. The transparency also includes extra answers that will not be used.

1. Simplify: $\dfrac{1}{3} + \dfrac{2}{5}$

    (Point Value: 5)

    Transparency Answer H; $\dfrac{11}{15}$

2. Using inequality or equality symbols, compare $\dfrac{8}{13}$ ■ $\dfrac{6}{13}$.
   (Point Value: 5)

    Transparency Answer B; >

3. Simplify: $\dfrac{2}{3} \times \dfrac{4}{5}$

    (Point Value: 5)

    Transparency Answer K; $\dfrac{8}{15}$

4. Simplify: $\dfrac{1}{3} \div \dfrac{5}{13}$

    (Point Value: 10)

    Transparency Answer E; $\dfrac{13}{15}$

5. Simplify: $\dfrac{3}{5} - \dfrac{2}{15}$

    (Point Value: 10)

    Transparency Answer A; $\dfrac{7}{15}$

6. Using inequality or equality symbols, compare $\dfrac{1}{3}$ ■ $\dfrac{9}{27}$.
   (Point Value: 10)

    Transparency Answer J; =

7. Using inequality or equality symbols, compare $-\dfrac{2}{3}$ ■ $-\dfrac{1}{10}$.
   (Point Value: 15)

    Transparency Answer F; <

8. Convert .4 into a fraction.
   (Point Value: 15)

   Transparency Answer C; $\frac{2}{5}$

9. Convert .2 into a fraction.
   (Point Value: 15)

   Transparency Answer M; $\frac{1}{5}$

10. Convert 75% into a fraction.
    (Point Value: 20)

    Transparency Answer N; $\frac{3}{4}$

11. Convert 60% into a fraction.
    (Point Value: 20)

    Transparency Answer D; $\frac{3}{5}$

12. Using inequality or equality symbols, compare 0.75 ◼ $\frac{3}{5}$.
    (Point Value: 20)

    Transparency Answer B; >

13. Using inequality or equality symbols, compare .6 ◼ 40%.
    (Point Value: 25)

    Transparency Answer B; >

14. Using inequality or equality symbols, compare 20% ◼ $\frac{1}{5}$.
    (Point Value: 25)

    Transparency Answer J; =

15. Using inequality or equality symbols, compare .015 ◼ .15.
    (Point Value: 25)

    Transparency Answer F; <

Beginning Algebra Inquiry Activities: Fractions Transparency

A. $\dfrac{7}{15}$ B. $>$ C. $\dfrac{2}{5}$

D. $\dfrac{3}{5}$ E. $\dfrac{13}{15}$ F. $<$

G. $\dfrac{4}{5}$ H. $\dfrac{11}{15}$ I. $\dfrac{1}{2}$

J. $=$ K. $\dfrac{8}{15}$ L. $\dfrac{10}{15}$

M. $\dfrac{1}{5}$ N. $\dfrac{3}{4}$ O. $\dfrac{1}{4}$

Beginning Algebra Inquiry Activities: Teacher's Key

Factors & Exponents: Some answers on the transparency will be used more than once. The transparency also includes extra answers that will not be used.

1. Simplify to a number: $(-3)^4$
   (Point Value: 5)

   Transparency Answer H; 81

2. What is the greatest common factor of 48 and 32?
   (Point Value: 5)

   Transparency Answer M; 16

3. What is the least common multiple of 6 and 12?
   (Point Value: 5)

   Transparency Answer A; 12

4. Simplify to a number: $-(3^4)$
   (Point Value: 10)

   Transparency Answer D; -81

5. Using inequality or equality symbols, compare $(3^3)^3$ ▋ 3^6.
   (Point Value: 10)

   Transparency Answer E; >

6. Simplify to a number: $3(2 + 1)^2$
   (Point Value: 10)

   Transparency Answer P; 27

7. Simplify to exponent form: $6x^2 \cdot 3x^6$
   (Point Value: 15)

   Transparency Answer L; $18x^8$

8. Using inequality or equality symbols, compare $5^3 \cdot 5^2$ ▇ 25^6.
   (Point Value: 15)

   Transparency Answer C; <

9. What is the least common multiple of 6 and 10?
   (Point Value: 15)

   Transparency Answer G; 30

10. Simplify to exponent form: $18x^{3 \cdot 4}$
    (Point Value: 20)

    Transparency Answer N; $18x^{12}$

11. Using inequality or equality symbols, compare 4^3 ▇ $3 \cdot 3^3$.
    (Point Value: 20)

    Transparency Answer C; <

12. What is the simplest form of $\dfrac{12df^2}{24dg}$?
    (Point Value: 20)

    Transparency Answer F; $\dfrac{f^2}{2g}$

13. What is the greatest common factor of 48 and 36?
    (Point Value: 25)

    Transparency Answer A; 12

14. What is the least common multiple of 5 and 12?
    (Point Value: 25)

    Transparency Answer B; 60

15. What is the greatest common factor of $6f^3g$ and $4f^2g$?
    (Point Value: 25)

    Transparency Answer J; $2f^2g$

Beginning Algebra Inquiry Activities: Factors & Exponents Transparency

A. 12

B. 60

C. <

D. -81

E. >

F. $\dfrac{f^2}{2g}$

G. 30

H. 81

I. -27

J. $2f^2g$

K. =

L. $18x^8$

M. 16

N. $18x^{12}$

O. -16

P. 27

Q. $18x^4$

R. -30

Reproducible Student Worksheets

- Variable Expressions
- Coordinate Plane
- Graphing
- Solving Equations & Inequalities
- Fractions
- Factors & Exponents

Name_____

Beginning Algebra Inquiry Activities:
Variable Expressions

Write the answers to the following questions in the space provided.

1. What is the variable expression for "7 less than x"?

2. Simplify: $15 + 20 \cdot 3$

3. Using inequality or equality symbols, compare 0 ▢ -100.

4. Simplify: $-48 \div 2$

5. What is the variable expression for the "product of 7 and n"?

6. Simplify: $-16 + -2$

7. Using inequality or equality symbols, compare -12 ▢ -9.

8. Simplify: 9 • 8 − 12 ÷ 2

9. Simplify: -1(-2)(-3)(-4)

10. Simplify: 5(8 - 4) ÷ 5 ÷ 2

11. What is the next number in this pattern 1, 1, 2, 3, 5, 8, 13, . . . ?

12. What is the variable expression for "7 times the quantity 4 plus x"?

13. Simplify: 14 − -4

14. Using inequality or equality symbols, compare |-4| ▪ |4|.

15. Solve: 15x − 2x, for x=3

Name_____

Beginning Algebra Inquiry Activities: Coordinate Plane

Write the answers to the following questions in the space provided.

1. What is formed when two number lines intersect at their zero points?

2. What is the term for two numbers that tell the location of a point in a coordinate plane?

3. In a coordinate pair, which axis is referenced first?

4. In which quadrant, are the coordinates both positive?

5. What is the number of the quadrant for an ordered pair with two negative numbers?

6. What is the quadrant for an ordered pair when the first number is positive and the second number is negative?

7. Connect the points in the order given. What is the shape of (2, 4), (2, -2), (-5, -2), (-5, 4)?

8. Connect the points in the order given. What is the shape of (2, -4), (7, -1), (4, 4), (-1, 1)?

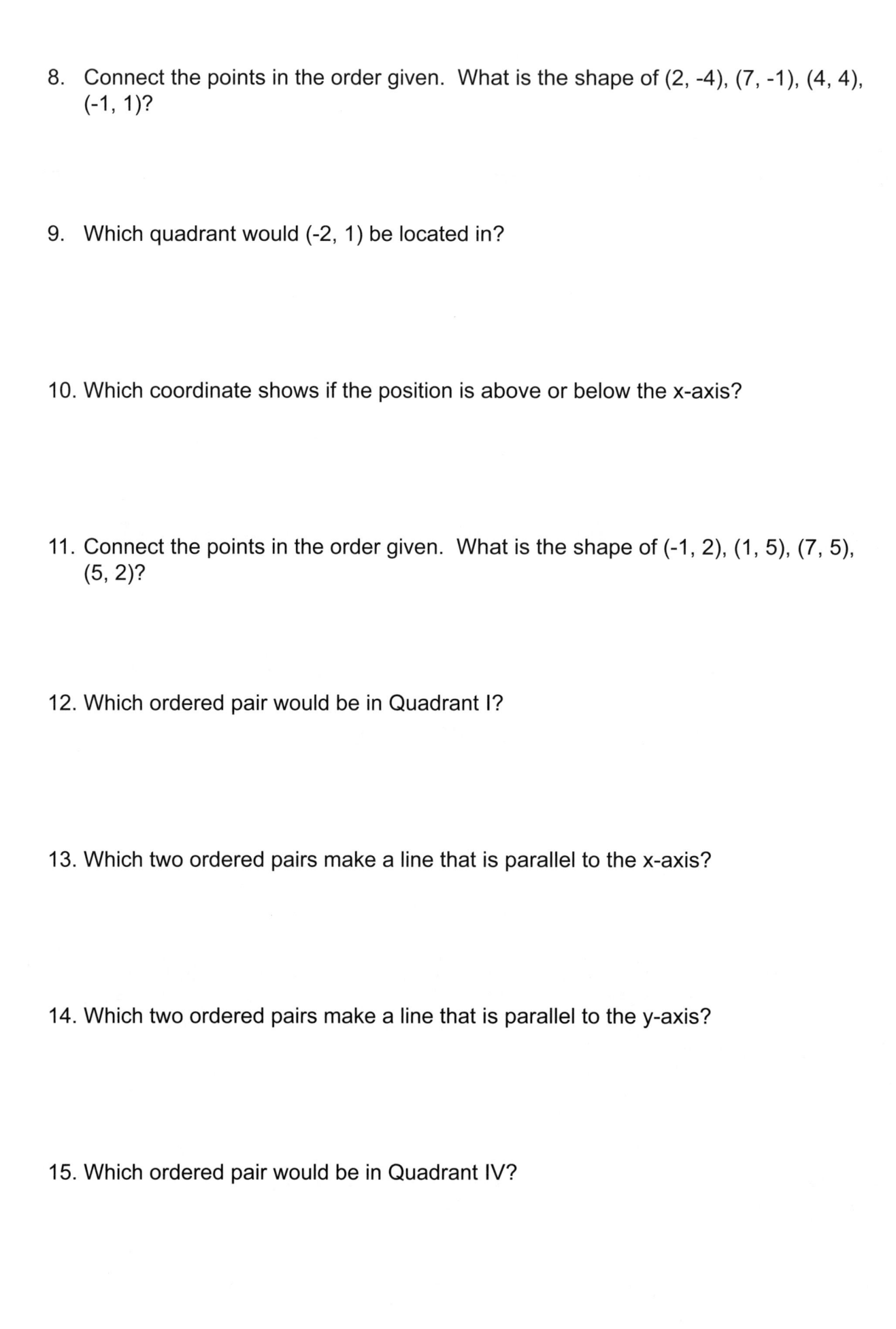

9. Which quadrant would (-2, 1) be located in?

10. Which coordinate shows if the position is above or below the x-axis?

11. Connect the points in the order given. What is the shape of (-1, 2), (1, 5), (7, 5), (5, 2)?

12. Which ordered pair would be in Quadrant I?

13. Which two ordered pairs make a line that is parallel to the x-axis?

14. Which two ordered pairs make a line that is parallel to the y-axis?

15. Which ordered pair would be in Quadrant IV?

Beginning Algebra Inquiry Activities: Graphing

Write the answers to the following questions in the space provided.

1. Which graph represents x = 2?

2. Which graph represents y = 2?

3. Which graph shows no correlation?

4. Which graph shows a line whose slope is 0?

5. Which graph shows a line whose slope is undefined?

6. Which graph demonstrates what happened to the weight of a group of students as the students increased in age?

7. Using the vertical-line test, which graph shows a relation that is not a function?

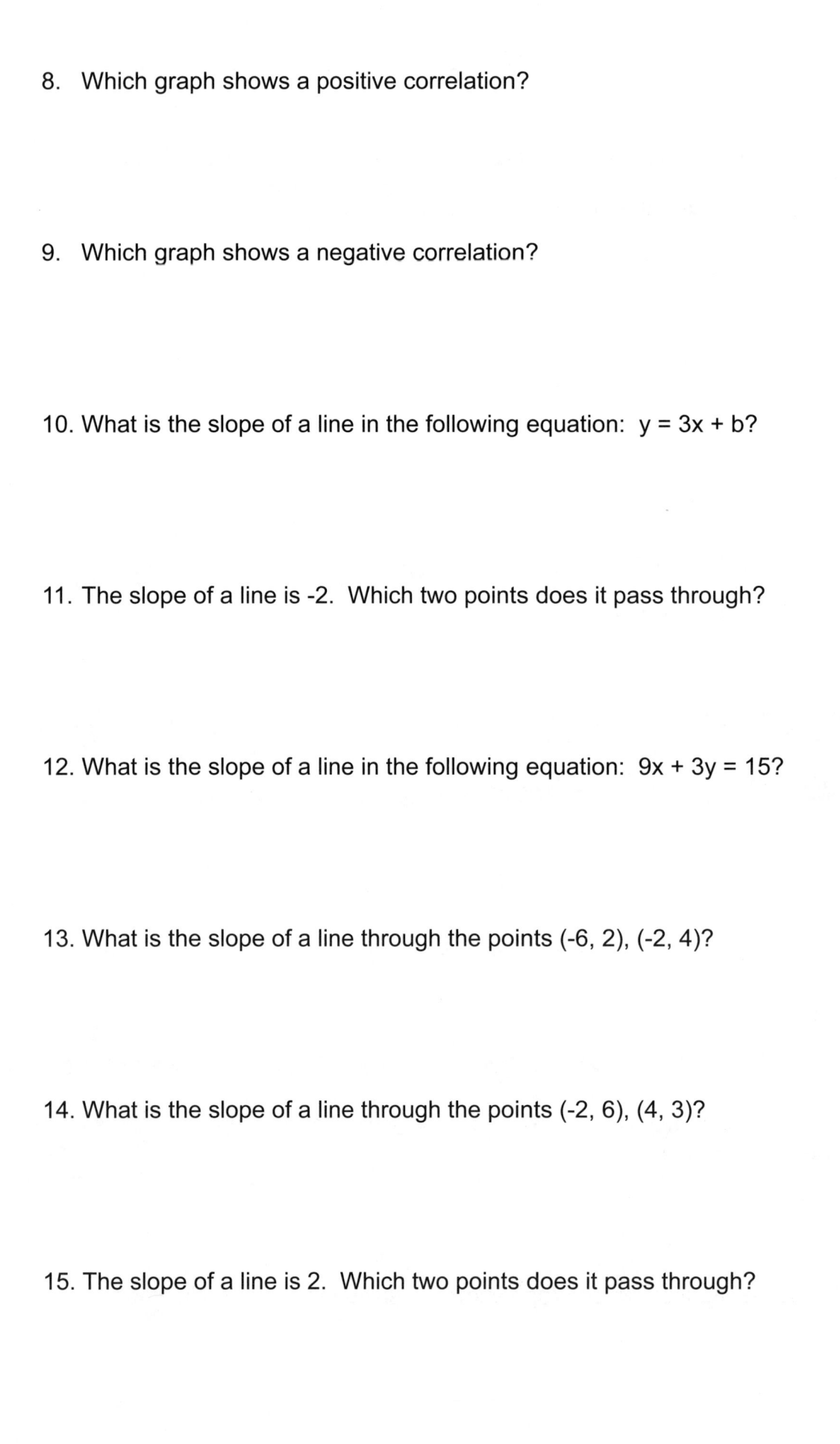

8. Which graph shows a positive correlation?

9. Which graph shows a negative correlation?

10. What is the slope of a line in the following equation: y = 3x + b?

11. The slope of a line is -2. Which two points does it pass through?

12. What is the slope of a line in the following equation: 9x + 3y = 15?

13. What is the slope of a line through the points (-6, 2), (-2, 4)?

14. What is the slope of a line through the points (-2, 6), (4, 3)?

15. The slope of a line is 2. Which two points does it pass through?

Name_____

Beginning Algebra Inquiry Activities: Solving Equations & Inequalities

Write the answers to the following questions in the space provided.

1. Which equation shows the Commutative Property of Multiplication?

2. Which equation shows the Distributive Property?

3. Which equation shows the Associative Property of Multiplication?

4. Which graph shows y < 3?

5. Which graph shows y = 3?

6. Which graph shows y ≤ 3?

7. Solve: -3x + 6 = 18

8. Which graph shows y > 3?

9. Which graph shows y ≥ 3?

10. Solve: x + 1 ≤ 5

11. Which graph shows the value of x in x(9 + 1) = 30?

12. Solve: $\frac{n}{2} - 8 = -6$

13. Which graph shows the value of x in 6(9 – 7) + x > 15?

14. Which graph shows the value of x in x – (9 – 2) ≤ -4?

15. Solve: 2y + 7 ≤ -1

Name_____

Beginning Algebra Inquiry Activities: Fractions

Write the answers to the following questions in the space provided.

1. Simplify: $\dfrac{1}{3} + \dfrac{2}{5}$

2. Using inequality or equality symbols, compare $\dfrac{8}{13}$ ▪ $\dfrac{6}{13}$.

3. Simplify: $\dfrac{2}{3} \times \dfrac{4}{5}$

4. Simplify: $\dfrac{1}{3} \div \dfrac{5}{13}$

5. Simplify: $\dfrac{3}{5} - \dfrac{2}{15}$

6. Using inequality or equality symbols, compare $\dfrac{1}{3}$ ▪ $\dfrac{9}{27}$.

7. Using inequality or equality symbols, compare $-\dfrac{2}{3}$ ▪ $-\dfrac{1}{10}$.

8. Convert .4 into a fraction.

9. Convert .2 into a fraction.

10. Convert 75% into a fraction.

11. Convert 60% into a fraction.

12. Using inequality or equality symbols, compare 0.75 ▊ $\frac{3}{5}$.

13. Using inequality or equality symbols, compare .6 ▊ 40%.

14. Using inequality or equality symbols, compare 20% ▊ $\frac{1}{5}$.

15. Using inequality or equality symbols, compare .015 ▊ .15.

Name_____

Beginning Algebra Inquiry Activities: Factors & Exponents

Write the answers to the following questions in the space provided.

1. Simplify to a number: $(-3)^4$

2. What is the greatest common factor of 48 and 32?

3. What is the least common multiple of 6 and 12?

4. Simplify to a number: $-(3^4)$

5. Using inequality or equality symbols, compare $(3^3)^3$ ▪ 3^6.

6. Simplify to a number: $3(2 + 1)^2$

7. Simplify to exponent form: $6x^2 \cdot 3x^6$

8. Using inequality or equality symbols, compare $5^3 \cdot 5^2$ ▮ 25^6.

9. What is the least common multiple of 6 and 10?

10. Simplify to exponent form: $18x^{3 \cdot 4}$

11. Using inequality or equality symbols, compare 4^3 ▮ $3 \cdot 3^3$.

12. What is the simplest form of $\dfrac{12df^2}{24dg}$?

13. What is the greatest common factor of 48 and 36?

14. What is the least common multiple of 5 and 12?

15. What is the greatest common factor of $6f^3g$ and $4f^2g$?